八卦虎爪雙刀
Advanced Tiger Claw Double Knives

Ancient Weapons of Kung Fu

八卦虎爪雙刀
Advanced Tiger Claw Double Knives

Ancient Weapons of Kung Fu

Paul Koh

KUNG FU IN A MINUTE PUBLISHING

一分鐘功夫

KUNG FU IN A MINUTE

KUNG FU IN A MINUTE PUBLICATIONS

佛光普照

Dedication

The study of Kung Fu weaponry is deep and wide. I have been so lucky to have a teacher that is an expert in Kung Fu weaponry, swords, knives, staff, spear, every esoteric and exotic weapon in the pantheon of ancient Chinese weaponry. I am astounded at his ability to not only play all these various weapons, but to articulate the movement of the weapon and draw its analogies to the empty hand in such a seamless way everything seems just to flow. His movement goes far beyond that of technical execution and becomes something more sublime, yet, at the same time, in the artistry, when he is enacting the movements of said weapon, my teacher seems to weave a tapestry of battle. It is clearly evident that every movement is being practically utilized and applied.

The exposure that I've received through his tutelage and guidance has had a profound impact on my understanding of Kung Fu weaponry and the study of Chinese martial arts overall. This in and of itself is no small feat. It is a true awakening and enlightenment not only to the practice of, but the still relevant purpose of, training in these ancient weapons during these modern times. The gratitude that I have cannot be adequately expressed or repaid. The study of Kung Fu as an art form and the weaponry that goes along with it under the guidance of a supreme master of the art is humbling. The only thing that I can even muster to say is, thank you, Sifu.

Table of Contents

Introduction

The introduction of weapons to the empty hand art of Kung Fu is always a point of contention. No one can quite put their finger on what was first, weaponry or empty hand in the use of combat. The art of Kung Fu stretches the entire length and breadth of Chinese history. Therefore, this point in time is lost in the midst of the eons gone by. Yet I wish to bring another point. Does it really matter at all?

And I say this because in this text, we explore the dual relationship proposed by both empty hand and weaponry training. Neither one presides over the other, yet both are echoed in this equal footing, one giving rise to the other while the other fosters. The double knife techniques exhibited in this text are closely related if not directly to all the techniques and movements of the Tiger Claw System. With or without, they still resound in their profound nature with the true spirit of the Chinese martial arts and the fighting prowess of the tiger.

The question in one's mind should not be which one came first or gave rise to the other, but rather to revel in and appreciate the knowledge and wisdom of both intertwined into one.

The Mysterious Origins of the Butterfly Knife

Within the Southern systems of Kung Fu, one of the first weapons to be learned is often the butterfly knife because of its close kinship and resemblance to the empty hand movements of the form. The butterfly knife is a clear, direct example of the relationship between empty hand training and weaponry. This makes it a natural outgrowth of what the student is learning. The stabbing, slashing and cutting with the weapon relates directly to various techniques of the empty hand—cuts, strikes, blocks and so on. These empty hand techniques can be readily converted for usage with the double knives. These short, stout, strong blades are an outgrowth of the movement of the practitioner and his techniques.

The true origins of theses knives are unknown, but most assuredly they are of a peasant beginning. There are many variations on the design, but the most common is a sturdy blade, wide and broad at the base and narrow towards the tip. Its construction seems very similar to a cleaver of sorts, so this may hint at its obvious original purpose. It is dull on the ridge and flat areas which allows for blocking, deflecting, intercepting and fielding the opponent's weapon, while the edge is as long as the forearm of the practitioner, which allows for great reach in all actions of cutting, hacking, chopping, slashing and stabbing.

The most unique feature of the construction of the double butterfly knives is undoubtedly its hand guard. The unusual aspect of the hand guard is its curved shape that protects the hand, or can be utilized to ensnare an opponent's weapon and also allows for the blade to be flipped 反刀 (fan do) into a reverse position, pressed against the forearm. This allows for slashing cuts to simulate close range elbow strikes as found in our Tiger Claw System.

Even though they are of humble origin, the ingenuity of the Southern masters to meld their knives together with this unique hand guard may have been inspired by European hand guarded swords. Secondary applications that are less lethal but utilize the hand guard would be in the concept and tradition of a brass knuckle where the outer hand guard simulates a punch or strike, as well as using the butt or the end of the blade for striking. In this way, we see the double knives are an extremely versatile and well-rounded everyday type of weapon.

These double swords are unique to southern China. They are shorter than a sword and longer than a dagger, easily carried and concealed, but extremely deadly. These knives

were not used on the battlefield, but rather by civilian martial art practitioners for their own self-protection. We see these knives grow in preference in the 1800s and 1900s and become a mainstay in many if not all Southern systems across the board.

In many skirmishes and battles between rival familial groups, clans and gang turf wars, be it in Hong Kong, China proper, or in US Chinatowns, the butterfly knives have seen extensive use and popularity. They were a weapon of choice for Tong members, merchant marines, mercenaries, civilian guards and local militia. The use of the double knives within Southern systems was so prevalent that it was reported in many encounters by Westerners as early as the 1820s.

The knives themselves have a true yin and yang nature to them, traditionally being called mother and son knives. The reasoning behind this is the mother and son relationship, being that one knife follows the other knife very closely as well as one knife supporting the other knife in its action. Sometimes the knives are paired tightly together. Sometimes they are in opposing hand position and direction. Simply said, when one is doing a defensive maneuver, the other one is attacking and vice versa.

Another term we can use is twin knives. They are akin to each other as twins. They respond and adapt to each other's actions and are closely woven together, inseparable in their application, two opposites being used simultaneously to create the one whole concept of the double knives. This aspect of having two knives hidden within one sheath at that point in time was unique and surprised many western onlookers and collectors. The butterfly knife/double knife became a sought after piece for many European and American weapon collectors.

The variety of positions that can be played with the butterfly knives is inexhaustible and therefore makes these paired weapons a real jewel in the crown of the Southern Kung

Fu weapons arsenal. The double butterfly knife is truly an extension of the individual practitioner and his knowledge of the basic tenets and principles of his system and helps to further accentuate and deepen his understanding of his Kung Fu training overall.

Despite the outward simplicity of these double knives, the profoundness of understanding that they imbue upon the practitioner is infinite in its scope. Just because they are common or popular does not lessen the variety of techniques and knowledge one can derive by playing these seemingly simple weapons. In this way, the butterfly knife is an indispensable tool in learning the art of Kung Fu.

Fundamental Movements of the Butterfly Knife

The butterfly knife, or double knives, is of paramount importance to all Southern systems of Kung Fu. It is one of the prominent weapons in their arsenal. The versatility and utility of this weapon cannot be overstated. Its practical usage and application speak directly to the mindset of the southern Kung Fu practitioner regardless of system.

Due to its structure, nature and concept, the butterfly knives go hand in hand with just about every physical gesture, movement and technique that can be found in the various Kung Fu systems of the southern tradition. The structure of the knife itself speaks directly to its practical application and usage.

Strong and sturdy, long enough but not too long, short enough but not too short, wider at the base and narrow at the tip, the butterfly knife is easily used for both offensive and defensive maneuvers simultaneously. It has the ability to be utilized fully extended, in reverse position or in myriad combinations. The butterfly knife was and is a go-to weapon for the Southern style practitioner.

The butterfly knife can be used in long range movement, middle range, as well as close, tight in-fighting when applied in reverse position against the forearm. In this way, the butterfly knife uniquely mimics the approach of the empty hand fighting concept of many of the Southern systems of Kung Fu. Based upon its sturdy construction, the butterfly knives have quintessential movements that all must learn and master.

Offensive and defensive maneuvers of the butterfly knives can be used separately and/or in combination with one another. These techniques include:

Slashing
Chopping
Stabbing
Slicing
Parrying
Deflecting
Blocking
Diverting
Hooking & Ensnaring

Hand and Finger Position for Holding Butterfly Knife

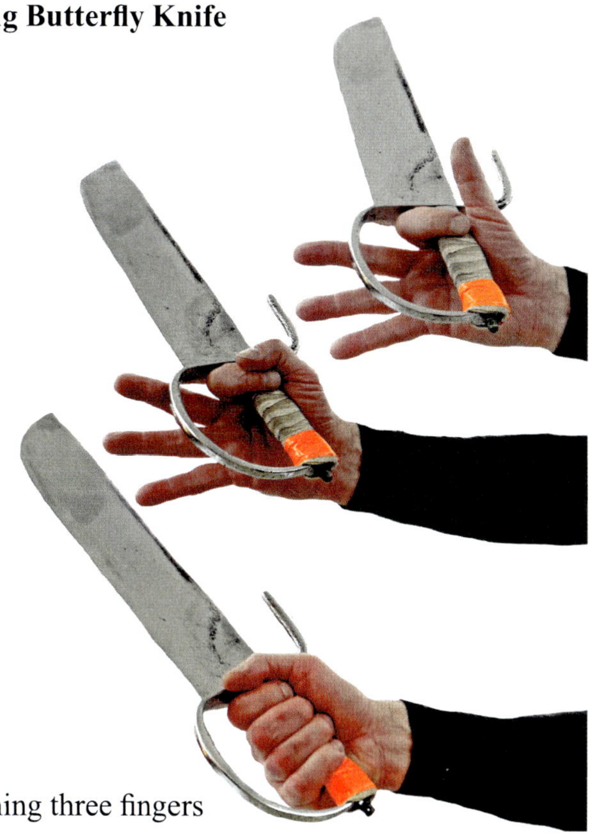

Handle of knife sits comfortably in palm of hand.
Thumb and index finger form a ring to hold knives while remaining three fingers lightly grip knife for manipulation and added dexterity.

Flipping Butterfly Knives

Hold butterfly knife in hand with index and thumb creating a ring.
Allow three remaining fingers to flip and fold butterfly knife in a downward positionwhile using hand and wrist to brace dull side of knife against forearm, index finger placed on handguard for extra support.
Reverse procedure to flip knife back upright.

八卦虎爪雙刀
ADVANCED TIGER CLAW DOUBLE KNIVES

Very few weapons can personify the essence of the Southern systems of Chinese Kung Fu and that of the action of the fighting tiger, as well as the double butterfly knives. The double butterfly knives are well-known throughout many, if not all of the Southern systems of Kung Fu, each system having its own unique rendition and application of these fearsome weapons.

This particular master set put together by my sifu, Master Tak Wah Eng, for me personally, is quite unique. It not only covers the gambit of all the typical techniques found within the double knife sequence, but uniquely intertwines and has blended with the characteristics of the tiger. This dynamic representation of the double butterfly knives and the tiger claw techniques is something special. From the beginning of this form, we can see the attributes of the silhouette of the tiger in every stroke, cut, chop and stab of these double knives. The ferocity and grace that is blended in so well smacks of the tiger and its prowess. Behind the usage of the double knives is the attacking tiger's attitude.

There are no defensive movements within this set. From the beginning, it is an onslaught of attacks, thrashing and slashing through the opponent at every turn. The double knives in this particular set have a penchant for not only encompassing the various ranges, but in particular, utilizing the reversed knife position buttressed up against the forearm of the practitioner to cut and stab in this unique position. The folded knives held tightly to the forearms and close to the body bring the individual practitioner to an up close and personal position with his attacker. Cuts are done in such a way that you don't see them coming. In a flash or in the blink of an eye, you're cut wide open. Both knives are intertwined together with the body of the practitioner making for a full onslaught of the tiger's attack. The unique aspect of extending and retracting the butterfly knives is seen in many of the other sets from different Southern families of Kung Fu, but in my personal opinion, not to the extent that it is being utilized here.

One of the most important and intriguing aspects of the double butterfly knife is that this double weapon requires the usage of not only both hands simultaneously, but both hands using two different weapons at the same time. If we look at the pantheon of weapons found within the Chinese martial arts, the majority of them are long and both hands will be required to wield a single weapon. In this instance, along with other weapons such as double straight swords, double sabers, double daggers and the like, the double butterfly knife will require each hand, left and right respectively, to manipulate a weapon on its own.

Unlike the others that were mentioned, the double butterfly knives will act in concert, but also more uniquely, separately. That is to say the left may not necessarily follow in sync with the right or vice versa. Each blade can mirror or be a twin to the other or do the exact opposite. One can be used for entrapping while the other one is cutting. One can be used for deflecting while the other one is slicing. This type of action will also be seen when using the double iron rulers. This forces the individual practitioner to raise their level of awareness, compensation of body movement, weight, and most particularly the dexterity required.

This heightened sense of awareness and dexterity with prolonged practice of the double knives will greatly enhance the student's ability not only to work weapons of this nature but also increase their empty-handed abilities. This is a hidden benefit that one can find within the butterfly knife form. One will also be able to increase the strength and integrity of the hand, wrist, fingers, and forearm, as well as the entire arm from constant manipulation and flipping of the butterfly knives back and forth. Of course, this will take quite some time, but it is a wonderful residual benefit that is gained. Overall, the 18 Tiger Butterfly Knife is a wonderful extension to the tiger claw techniques within the system and will greatly benefit the student that is diligent and dedicated in his practice.

Each section of this master form is, in actuality, a form unto itself and can be played as such. The entirety of the form is incredibly long and difficult for anyone other than a very advanced practitioner to attempt. The overall flavor of this set is uniquely tiger claw and must be approached with this mindset and a deep understanding of the techniques of the Tiger Claw Kung Fu System. Without this as a background, the pure mimicry of the movements will fall far short of their true intention. Therefore, a good, sound, basic understanding of the dynamic approach of the tiger and the Tiger Claw System must be at the core for the practitioner to make any headway in this set.

Opening Salutation

The beginning of the form dictates how it will be executed from the salutation on. The knives are already held in a reverse position in the left hand, hidden and obstructed from all. Being held in this way, the knives are open and ready to cut from the first action, unlike in other sets where the open blade is in a neutral position. The first movement is already seen as aggressive and slashing out. From that point on, there is no holding back. Every flip, every turn of the knives is meant to destroy the adversary, slicing them open at every turn. The salutation itself is courteous, bowing to the left, right and center, but with the blades facing outward, still gives the meaning and connotation that we are not here to dance, but rather to fight to the death.

First Section

In this section, we open with brushing the blades. The hacking, chopping, slicing to left and right sides and drawing back, brandishing the blades almost welcomes the next opponent.

Second Section

The second section of the form will begin with the practitioner in the classical tiger stance with both knives forward, portraying the claws of the tiger, cutting forward then up immediately. The blades will be flipped over, braced against the forearms, using them in every onslaught and twist and turn. We see the blades connected with the body, portraying the claws of the tiger, slashing and ripping, as well as cutting outward with many specific and deadly elbow techniques only to recoil back and present the blades as though they were the fangs of the tiger.

Third Section

This section continues with the blades fully extended but in short, tight upright traps and cuts, highs and lows, smoothly transitioning into wider open movements with the full blade extended, up, down, out, in and around, cutting, stabbing, extending to the opponent's front and back, over as well as under from front to back and left to right. Everyone that enters into the circle of this section will be cut. Following this full extension piece, the practitioner flips the blades back onto the forearm, displaying the classical double and single tiger claw techniques. A variety of stances are used, covering all the areas that radiate out from the double knife practitioner.

Fourth Section

Centralizing once more into the horse stance, we see the knives come out and in, cutting as though they were a giant scissor, transitioning from right to left, panning the face, blocking the entire body up and down, slashing outward to cut at the waist and employing various traps and disarming techniques to stab and thrust at the opponent with the blade's full extension.

Fifth Section

Assuming the position of an attacking tiger with both knives thrust forward, we again flip the knives to be braced on the forearms and emulate forearm and claw techniques of the tiger, trapping and slashing, deflecting and striking. Each elbow cut is tightly integrated with every action, stance and body movement. There seems to be no knife at all, just the action of the tiger. The butterfly knives now have taken the place of the claw hand formation of the tiger claw practitioner. Rising, spinning and twisting with the knives is prominent throughout this section.

Sixth Section

In part six, we continue and delve deeper into utilizing the butterfly knives in the reverse position. Finding ourselves playing in even tighter intertwining circles while utilizing trapping and striking movements, as well as sweeping kicks and elusive footwork.

Seventh Section

Part seven brings us back to full extension, using the double knives to slash and cut, stab and entrap the opponent's weapon.

Eighth Section

Segueing smoothly into part eight, we again collapse the blades inward, utilizing cover and strike into spinning elbow techniques with the blades in this hidden position. We draw the opponent in, striking and cutting down, up, left and right, leaving no space for them to run. The double blades still being on the offensive, we see the form direct us to cut down and to the side, also striking out with the pommel of the blade, finally culminating this section in opening the blades and brandishing them for the next section to begin.

Ninth Section

Part nine, the final chapter in the 18 Tiger Butterfly Knife Set, is a smooth and unique blend of a variety of double knife techniques utilizing all aspects of the double knives, including reverse position to ensnare, strike, chop and cut the opponent with a display patterned after the double and single tiger claw techniques as well as tiger fist strike facing to the front. As we segue and turn to the back and open the blades, the double knives emulate longer hand movements such as swinging punches, uppercuts and the like, to return back to the front, coming full circle and drawing the tiger back into its lair with a traditional tiger claw salutation.

Note: Hold both knives together in inverted position, pressed to forearm and held with thumb and index finger.

Step forward with right cross stance, bringing knives and hand to right hip in clockwise motion.

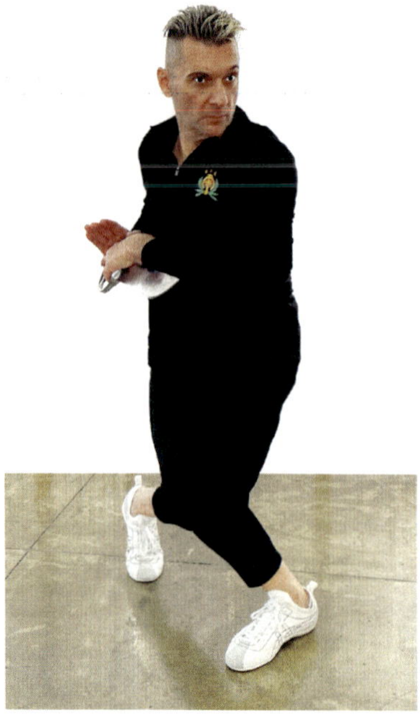

Step forward into left tiger stance, presenting salutation, knives in front of right hand, both crossing at wrist.

Step forward to left cross stance, bringing knives and right hand over to left hip in counterclockwise motion.

Step back three steps into left tiger stance, opening up both knives in counterclockwise downward slashing motion and assuming ready position.

Bring both knives inward and up in slashing fashion, drawing both forward in front of body in V-shaped formation, assuming half-horse position.

Turn to left side, brushing downward with both knives, stepping into left front stance, chopping downward. Retract knives, slicing back, assuming left tiger stance position.

Turn to right side, brushing downward with both knives, stepping into right front stance, chopping downward. Retract knives, slicing back, assuming right tiger stance position.

Bring both knives inward and up in slashing fashion, drawing both forward in front of body in V-shaped formation, assuming half-horse position.

Invert both knives, flipping inward, bracing onto either forearm.

Step into right cross stance, blocking with right hand up.

Step towards left with left cross stance, left knife deflecting.

Continue stepping to left with right cross stance with right knife deflecting.

Uncoil legs to shift into left reverse bow stance, thrusting backward with left hand, cutting downward with right to back right corner.

Realign right leg to center in horse stance, cutting forward and back with both knives.

Step with right foot to 45 degree angle, slicing down and thrusting back in reverse right bow stance facing front left corner.

Aggressively move forwad with three consecutive reverse knife elbow strikes, right, left, right.

Step back three steps while bringing both knives inward and up in slashing fashion, drawing both forward in front of body in V-shaped formation, assuming half-horse position.

Circle knives downward to front left corner, trapping with left knife, chopping high with right, then immediately chopping low with left. Shift body to face front right corner in half-horse with knives in high and low guard position. Chop high with left, then immediately chop low with right. Shift body to face front left corner in half-horse with knives in high and low guard position.

Bring knives forward to center, slashing upward and around, stepping into left bow stance with double under cut, facing directly forward.

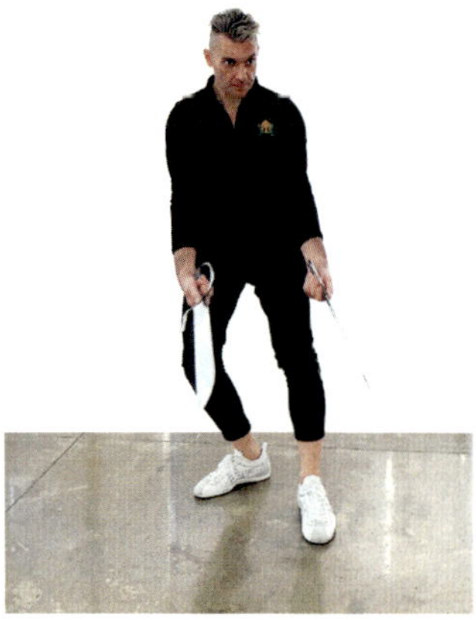

Draw knives downward, raising up to head, chopping downward parallel to floor.

In left bow stance, turn both knives outward to either side.

Turn waist to assume left horse stance, folding both knives inward, right on top of left.

Shift back to left bow, cutting forward with both knives simultaneously.

Circle both knives in clockwise fashion, slashing down with left, stabbing downward with right on a 45 degree angle.

Step back with left foot into right cross stance, chopping front and back with both knives.

Step out with left foot into side stance, circling left knife counterclockwise, slashing diagonally with right.

Step to right side with left foot into right cross stance, cutting upward with right knife.

Turn to face left side. Draw both knives downward, raising them up into double stab in left front stance.

Turn to look towards front, extending both knives outward, chopping front and back.

Shift body to right side stance, extending both knives out to cut under, front and back.

Draw knives inward, brushing the face counterclockwise. Step forward into right horse and cut parallel to floor.

Draw right foot in, left foot back, exchanging position, assuming right bow stance with high X-block facing back.

Draw knives inward, brushing the face counterclockwise. Step forward into right horse and cut parallel to floor.

Fold both knives inward. Step up to front right corner, deflecting down with left knife and cutting up with right in left bow stance.

Drop right knife downward, spinning body around to left front corner, slashing up with left knife in tiger stance, drawing right knife to waist.

Shift backwards to face front, cutting horizontally with right hand, vertically with left and then again horizontal right elbow slash.

Open knives back to original position. Block down and around, raising left leg, stepping into left bow, chopping down with both knives.

Retract body into single-leg stance, executing three fast triple chops. Immediately, step into left horse stance thrusting forward with left knife, right in guard position.

Draw right leg around and back to face back with double parallel chop in half-horse.

Fold both knives inward. Step back and turn to front, executing right, then left, right elbows sequentially, traveling forward in tiger stance.

Turn to back left corner in kneeling position with side elbow slash.

Turn to front right corner with rising elbow slash.

Come back to center horse stance facing front, circling knives downward and then inward, executing scissor cut, ending with both knives on either side.

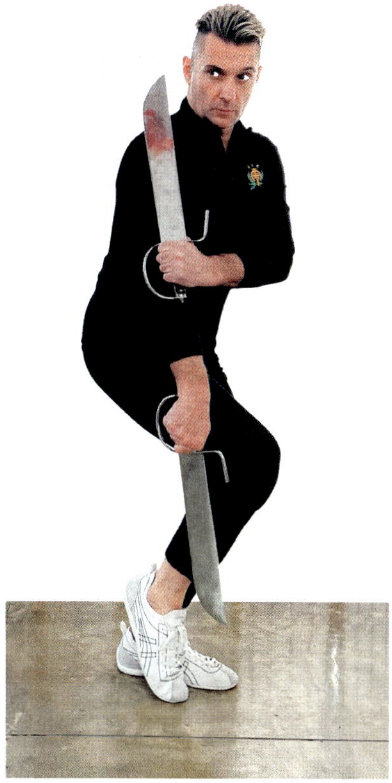

Execute three inward circling cuts, left, right, left. Assume right toe stance with right downward chop and high left guard hand, facing right side.

Execute three inward circling cuts, right, left, right. Assume left toe stance with left downward chop and high right guard hand, facing left side.

Step into left side stance facing front, drawing left knife back to waist, cutting outward with right at 45 degree angle.

Reassume horse, placing right knife over left in X-position.

Rotate knives back to side of body, assuming left tiger stance.

Step forward into left bow, thrusting both knives out in low X stabbing position.

Bring both knives inward and up in slashing fashion, drawing both forward in front of body in V-shaped formation, assuming half-horse position.

Invert both knives, flipping inward, bracing onto either forearm.

Step forward into right bow, blocking high and low.

Shift body to side stance. Deflect with left knife high. Follow through with right knife cut. Draw left knife back and extend out with right.

Shift to right side stance, executing left diagonal knife down.

Step forward into right horse, cutting in with right knife.

Shift back to right side stance, cutting diagonally upward with left knife.

Turn body to face front right corner. Open both knives to front and back, looking at left front corner.

Fold knives inward, executing low sweeping kick to front right corner. Spin around to face back right corner, with thrusting strike of right pommell of blade.

Shift backward, executing left slicing cut.

Open right knife, slashing upward in horse, facing back left corner.

Continue and step into low right cross, cutting from under.

Step up to front right corner in left bow, deflect with left and slashing up with right.

Slide left foot to left front corner, simultaneously folding right knife in, assuming X-position in front of torso.

Step forward with right foot into half-horse, executing high block facing front.

Shift to right side stance, executing diagonal left elbow cut downward.

Immediately open again with right hand, executing high diagonal left slash.

Open knives and bring both inward and up in slashing fashion, drawing both forward in front of body in V-shaped formation, assuming half-horse.

Fold knives and turn them inward, deflecting high and low. Immediately block out with right, striking with pommel of left in right bow stance facing front right corner.

Rotate knives in front of body, turning to face front left corner, ending in high and low cutting strike.

Open both knives to front and back, looking at right front corner.

Execute right upward slashing cut, followed immediately with downward left slashing cut, traveling forward.

Execute left upward slashing cut, followed immediately with downward right slashing cut, traveling forward.

Step back and turn body to the side, cutting horizontally with both knives.

Skip forward.

Right kneeling stance with X-block to the ground. Rise up, executing high X-block.

Circle knives overhead, stepping into left bow, chopping downward. Shift back into reverse bow, chopping down.

Circle knives overhead, shifting into right bow, chopping downward. Shift back into reverse bow, chopping down.

Fold knives inward, trapping with left, striking with right while stepping to front left corner.

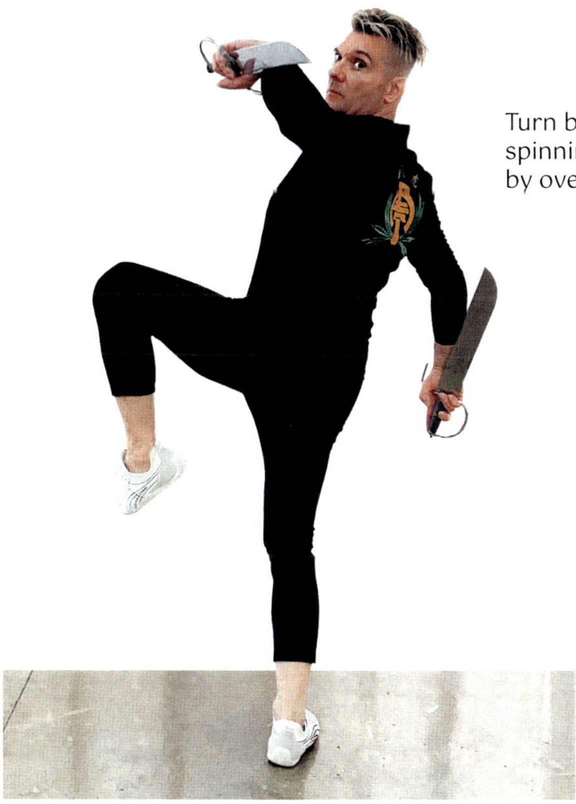

Turn body around to execute left spinning elbow, followed up immediately by overhand right strike.

Circle both knives in towards body, shifting into horse stance. Draw right back and thrust left forward.

Thrust downward with right, stepping into left side stance. Shift to right side stance, cutting back with left.

Deflect upward with right hand in modified horse. Immediately slash up with left in right side stance.

Turn torso to left side and execute slashing elbow.

Drop down into right kneeling stance, blocking high and low with both knives facing left side.

Slide into left side stance, cutting downward with right knife facing front.

Interlock right foot behind left, panning face with both knives, circling inward. Step forward into right horse stance, striking out with right knife, blocking with left.

Move into right bow, drawing right knife back, striking forward with pommel of left.

Step back with left foot into right cross, opening both knives, chopping front and back simultaneously.

Step forward with left foot into tiger stance, intersecting both knives in front of torso.

Fold knives inward, bracing to forearms.

Advance forward, simultaneously covering inward with right, left and right, ending up in left tiger stance, striking with right knife to front left corner.

Step to left with right horse, blocking torso with both knives. Immediately chop horizontally with right knife.

Step to right side with right foot into cross stance, circling both knives in clockwise rotation, first blocking with right, then striking out with left.

Shift body into left side stance, striking downward with right knife, facing front.

Cover and cut to left front corner.

Shift to other side, executing same.

Shift to left side stance, moving up slightly, executing criss-cross cut downward with both knives.

Retract back into left tiger stance. Slide up with left foot into bow stance, executing double pommel strike with knives.

Bring right foot behind, turning around to face back, opening knives and then intersecting them in front of torso in ready position.

Step forward into right horse, covering with left knife, chopping with right.

Turn body to face front, deflecting high with right, cutting outward with left.

Shift left leg slightly to front left corner, slicing downward.

Shift to right bow to right front corner, slicing upward.

Step directly forward with left foot into bow stance, executing center cut to throat.

Step back with left foot into right cross stance, chopping front and back with both knives.

Step forward with left foot into tiger stance, intersecting both knives in front of torso.

Fold knives inward, bracing to forearms.

Shift to right side, blocking with both knives. Shift to left, repeating same.

Open both knives, executing upward slashing action, drawing both knives to left side.

Fold both knives into crade of left arm. Draw knives and claw backward, simultaneously stepping forward with right foot into cross stance.

Step forward into left tiger stance, executing tiger claw salutation with knives and right hand to end set.

雙刀對棍
Butterfly Knife vs. Staff

The original concept of the development of weapons such as we find in the art of Kung Fu are, simply put, to better enable the warriors to extend their physical abilities of the body to more easily and effectively destroy the opponent. Therefore, we see that the weapons, in this case the Double Butterfly Knives and Staff, are far more than mere tools that one can arbitrarily pick up and wield in wild thrashing, but rather require a high degree of skill and precision. This skill and precision extends not from the weapon, but rather from the player himself and his deep-rooted understanding of the basic tenets of his chosen Kung Fu system. In this case, we speak of the Tiger Claw Kung Fu System.

Every step, every twist, every turn, all the techniques found within this matching set stem from this deep knowledge, understanding and of course innovativeness that is given to us by the countless hours of study and practice put into the Tiger Claw System of Kung Fu. As always stated by my Grandmaster Wai Hong, "We should be able to pick up an object and it becomes a weapon." This statement alone should convey the understanding that regardless of the weapon or tool, the tiger claw practitioner should be able to adapt his empty-handed physical training to envelop and utilize anything that is placed in his hands.

If we pay close attention to every position and technique displayed within this coupled matching set, we will see the attributes of the tiger and all the splendor of its techniques from the basic footwork and stances to the attacking demeanor. Both sides of the matching set are equal in the sense that we are looking at a mirrored image on either side. The knives match the staff; the staff matches the knives, both displaying the visage of the tiger's technique and ferocity.

Hidden within the weapon matching set are the techniques found within the Tiger Claw System and one could go as far as to say without much augmentation, we could play either way, with weapons or without, or even, if possible, substitute other various weapons to take the place of the two displayed here. This is the deep Kung Fu that is within the system and its ability to adjust and adapt to whatever the situation or tool put in one's hand.

Both the double knives and the staff are evenly matched. That is to say that neither side is winning or losing in this encounter but rather fostering and helping one another see their own individual attributes and capabilities. If one studies and practices long enough, one will be able to extrapolate and develop even more techniques that are seen immediately and on the surface.

In full display previously presented in the double knife form, we see all the variety of cutting, chopping, thrusting and trapping, deflecting techniques of the knives, as well as the mixture of fully extended double knife techniques and those in reverse position braced upon the forearm, mixing and matching as we go in a very intense high-speed matching set that if played incorrectly, either side, double knives or staff, could suffer injury. Therefore, it goes without saying that when learning this matching set, one must take caution and move slowly. The practice of empty hand matching sets is just as dangerous, but yet with one wrong movement with the knife or the staff, we may have a serious situation that one does not want to develop. So I caution the partners to go slowly and learn each movement intimately before increasing any power and speed in these techniques.

As aforementioned, neither side is winning. Therefore, if you strive to learn the side of the double-headed staff, you are not at a loss. It is a true and wonderful companion to the double knives and has been paired against the double knives traditionally. Both of these weapons help to foster the idea of extending the limbs, power, speed, and awareness of the body and the mind of the practitioner without regard to if they are playing one side or the other.

One should not feel that one weapon is more inferior than the other. This is not the case. The inferiority or supremacy of the weapon is solely based upon the player. This matching set is a cross-section of a very lengthy advanced matching set that by itself would require another text to display, so I'm only giving a smaller cross-section that will show the dexterity involved in either side. Again, more is not always more and less is not less. The depth of understanding of the practitioner is more important than the length of any particular matching set or the weapon that is being played, understanding that techniques and matching sets per se will change and evolve as played continually.

The pictures displayed here only capture what can be caught in the camera. The actual interaction of the two participants will vary and change over time and develop. The art of Kung Fu and fighting is not a one- or two-dimensional learning but rather three-dimensional. Therefore, do not feel that if you have not copied the picture 100% that you are wrong. Every individual practitioner has his own attributes, speed, and strength. In fact, the physical size of the individual, be it smaller, taller, and so on will impact and change the technique. This is the adaptability and innovativeness of real Kung Fu rather than sticking fanatically to one dogmatic concept and not being able to change or roll with the punches, as they say.

Prior to engaging in the matching set, it is strongly suggested to view the separate sections A and B respectively to understand the individual weapons themselves. These separate sections, if wished, can also be played as a form or weapon set unto itself before intertwining in the engagement with the partner. The partner is the other half of the equation and must not be abused or looked upon in this instance as the enemy, but actually as the sharpening stone through which we will be able to glean an edge for our techniques. Abusing or wishing to overwhelm the partner will only nullify and stunt either side's growth.

Once proficiency is attained with either side, then the practitioners may be able to extract one or two techniques and develop a shorter routine that can be used for direct application. It must be understood that one good cut from the knife or one strong strike from the staff should and would be able to end an encounter with a true opponent. This is not the object of the matching set but can lead to a greater understanding on how to develop techniques for application against an opponent. The basis of the matching set is to help the individual learn about themselves, about their weapon, about circumstances and scenarios, and how to adapt and grow.

Both partners assume ready position, double knives held forward in front of body, chest level, staff held forward at chest level.

DOUBLE KNIVES folds knives and retracts back into low left tiger stance, ducking under.

STAFF attacks high with right slashing strike.

DOUBLE KNIVES evades with left single leg stance.

STAFF follows up with low right sweep.

DOUBLE KNIVES opens both knives and attacks with high double chop, stepping forward in right bow stance.

STAFF blocks high with center of staff.

STAFF presses knives over and down to right side. DOUBLE KNIVES rolls with deflection, folding knives. Both partners spin around into hanging leg position.

Both partners step into right bow, mutually attacking one another.

DOUBLE KNIVES uses left knife to brush away staff, executing left leg hooking kick.

STAFF evades by raising right leg.

Both partners spin around 360 degrees. DOUBLE KNIVES drops down to right kneeling stance, attacking with right knife open.

STAFF executes low stab with left hand.

STAFF executes right sweep to deflect attack.
DOUBLE KNIVES jumps back.

DOUBLE KNIVES executes high X-block with both knives open in right bow stance.

STAFF steps into right bow stance with left downward strike.

DOUBLE KNIVES deflects staff with left knife, stepping to 45 degree angle, thrusting right knife.

STAFF shifts back with right foot to 45 degree angle, deflecting with right block.

DOUBLE KNIVES steps up, thrusting left knife.

STAFF shifts back, deflecting with left.

DOUBLE KNIVES steps up, thrusting right knife.

STAFF shifts back, deflecting with staff.

DOUBLE KNIVES steps up, stabbing with left knife.

STAFF parries, shifting backward with left side of stick.

DOUBLE KNIVES shifts to right, fielding strike with both knives folded.

STAFF turns around, executing extended strike with right side of staff.

DOUBLE KNIVES deflects staff upward with left knife, dropping down and opening right knife, executing chopping strike to legs.

STAFF raises weapon overhead and jumps over slashing cut.

DOUBLE KNIVES steps into right side stance folding right knife and deflecting with left.

STAFF lands after jumping and shifts into reverse cross stance, executing thrusting staff attack.

DOUBLE KNIVES shifts body to left, slashing with right knife.

STAFF drops down to avoid cut.

DOUBLE KNIVES opens both knives and attacks with double cut in right single leg stance.

STAFF rises up, blocking attack with middle of staff.

DOUBLE KNIVES spins to left, executing horizontal strike.

STAFF shifts to left side stance, executing downward right strike to block.

DOUBLE KNIVES brings knives in, using right to deflect, preparing to spin around.

DOUBLE KNIVES completes rotation,
attacking to groin in reverse bow stance.

STAFF skips back into right bow stance,
deflecting downward with middle section of staff.

DOUBLE KNIVES fold both knives and deflects upward with left.

STAFF steps forward into left bow, crashing down with right strike.

DOUBLE KNIVES shifts into low horse,
blocking down with right.

STAFF shifts downward into low horse,
attacking with left strike.

DOUBLE KNIVES evades sweeping strike, assuming left single leg stance.

STAFF cuts across with sweeping strike to legs.

DOUBLE KNIVES drops down to low horse with knives behind, ducking under staff strike.

STAFF shifts to right side stance, executing cutting strike with left side of staff.

DOUBLE KNIVES opens right knife, executing spinning step into cross stance, right knife chopping outward.

STAFF shifts back into right tiger stance, blocking with center of weapon.

DOUBLE KNIVES folds right knife and executes high elbow with left.

STAFF shifts to low left reverse bow to evade attack.

DOUBLE KNIVES steps down into left side stance, striking down right pommell of right knife.

STAFF shifts slightly, retracting staff to evade downward strike.

DOUBLE KNIVES raises right leg, opening right knife, stepping away from STAFF.

STAFF turns away from DOUBLE KNIVES, rotating staff downward with right hand.

DOUBLE KNIVES spins around, opening other knife in left single leg stance.

STAFF continues downward staff flower.

DOUBLE KNIVES steps back, drawing left leg up, brandishing both knives towards STAFF.

STAFF continues flower rotation, spinning around to assume left kneeling stance, brandishing staff upward towards DOUBLE KNIVES.

DOUBLE KNIVES steps away from partner into right bow, folding knives and deflecting outward with both.

STAFF steps away from partner into right bow, striking high with left hand.

DOUBLE KNIVES steps forward with left, drawing knives inward, slashing with right.

STAFF steps up to left, executing high right strike.

DOUBLE KNIVES shifts into left side stance, cutting outward with right knife.

STAFF thrusts out staff with right hand.

DOUBLE KNIVES shifts to right side stance, slashing with left knife.

STAFF shifts to right side stance, striking down to ground with left.

DOUBLE KNIVES turns inward, facing partner back to back with mid and high guard.

STAFF turns inward, facing partner back to back with mid-level lateral strike.

DOUBLE KNIVES shifts to right side stance, slashing inward with left to face partner.

STAFF shifts to right side stance to face partner, executing upward strike parallel to ground.

DOUBLE KNIVES opens both knives, encircling head and shoulders, spinning around towards left.

STAFF executes downward staff flower, spinning towards left.

DOUBLE KNIVES completes turn, extending both knives outward, then intersects both knives in front of body.

STAFF completes flower and raises staff overhead, placing it onto left shoulder with right tiger claw forward.

Both partners assume ready position with DOUBLE KNIVES folding knives inward.

DOUBLE KNIVES steps into left cross stance, covering body with both knives.

STAFF steps into left cross stance, covering body with right claw to shoulder.

DOUBLE KNIVES steps into right cross stance, covering body with both knives.

STAFF steps into right cross stance, drawing claw to waist.

Both partners spin around.

DOUBLE KNIVES prepares to chop in single leg stance.

STAFF completes rotation, preparing to strike.

DOUBLE KNIVES completes rotation, assuming left horse, double knife striking towards center.

STAFF completes rotation, assuming right bow stance, executing lateral strike.

DOUBLE KNIVES pushes staff away then executes simultaneous cut to head.

STAFF ducks under to evade.

DOUBLE KNIVES completes rotation and steps into left reverse cross stance with left chop.

STAFF rises, executing mid-block in left side stance position.

DOUBLE KNIVES spins away after kick towards right.

STAFF kicks away attacking knife with inside crescent kick to DOUBLE KNIVES' left arm.

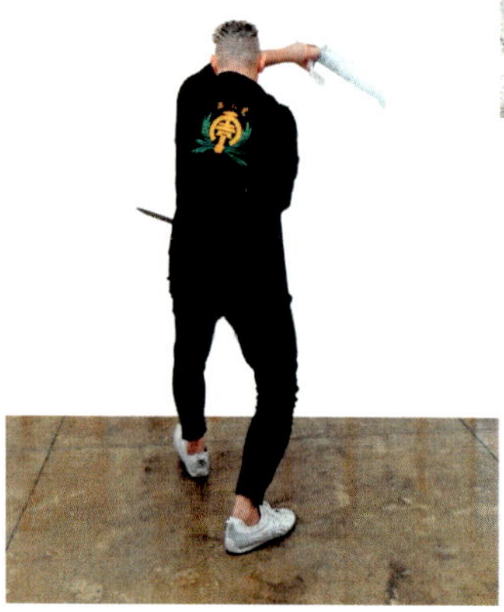

DOUBLE KNIVES completes spin, attacking with left stabbing knife.

STAFF spins around from crescent kick and blocks to center.

DOUBLE KNIVES attacks with
right knife stab.

STAFF strikes down with right hand to
block knife thrust.

DOUBLE KNIVES ducks down to avoid strike.

STAFF slashes overhead with left.

DOUBLE KNIVES rises up, using X-block.

STAFF strikes overhead with right.

DOUBLE KNIVES evades with single leg stance.

STAFF strikes upward with left to open knives and spins around.

DOUBLE KNIVES folds knives inward, trapping staff strike.

STAFF completes rotation, executing thrusting right strike in horse.

DOUBLE KNIVES slashes out with left.

STAFF ducks under.

DOUBLE KNIVES opens right knife, chopping to center.

STAFF rises to block.

DOUBLE KNIVES opens left knife, cutting low to leg.

STAFF evades in single leg stance.

DOUBLE KNIVES spins away from partner towards left, folding left knife.

STAFF executes staff flower, stepping away from partner.

DOUBLE KNIVES assumes half-horse with left knife folded, right knife poised for attack.

STAFF crosses weapon in front of body, ready for next encounter.

DOUBLE KNIVES folds right knife inward and begins to criss-cross step, encountering partner.

STAFF crosses over to criss-cross partner with downward staff flower.

DOUBLE KNIVES rotates knives around face and body.

STAFF completes downward flower, turning around to prepare low extended strike with staff.

DOUBLE KNIVES extends both knives forward and thrusts out with right toe kick.

STAFF raises left leg up and extends staff strike to ground.

DOUBLE KNIVES steps down into reverse right cross stance, striking down with right knife.

STAFF retracts weapon, throwing it up in air, catching it from behind. Both partners move around in each other.

DOUBLE KNIVES spins around to face partner, cutting in and down with both knives.

STAFF rolls weapon off back, crossing over with right foot, executing downward staff flower.

DOUBLE KNIVES entwines both knives around head and spins towards right with left inner hooking kick.

STAFF completes flower and turns around to face partner.

Having completed change of positions, both partners face each other at the ready.

DOUBLE KNIVES blocks upward with left knife.

STAFF steps forward with right foot and overhand strikc.

DOUBLE KNIVES uses right knife to press strike downward.

STAFF follows through with knife press.

DOUBLE KNIVES turns towards left, executing left elbow strike.

STAFF shifts over to left side, blocking knife thrust.

DOUBLE KNIVES evades sweep by raising left leg.

STAFF immediately follows up with right sweep to leg.

DOUBLE KNIVES spins to right, executing right elbow strike.

STAFF shifts to right bow, using center of staff to block.

DOUBLE KNIVES cuts to STAFF's head then draws back into X.

STAFF ducks under, evading slash.

DOUBLE KNIVES shifts back
into left tiger stance.

STAFF rises up.

DOUBLE KNIVES executes overhead inward crescent kick.

STAFF shifts into low back stance to evade kick.

DOUBLE KNIVES deflects outward with right knife.

STAFF rises up, moving around partner, and steps forward, executing downward strike.

DOUBLE KNIVES shifts back into horse, blocking downward with left knife.

STAFF shifts to left horse, striking upward towards groin.

DOUBLE KNIVES spins towards right, executing right tiger tail kick.

STAFF shifts to left side, deflecting kick with center of staff.

DOUBLE KNIVES steps into left bow stance, cutting down with right knife, preparing to yank staff down with hand guard of knife.

STAFF raises weapon to block high.

DOUBLE KNIVES shifts into left horse, executing horizontal left knife slash.

STAFF rotates staff over, using center to deflect.

DOUBLE KNIVES blocks upward, crossing both knives.

STAFF executes high left strike.

DOUBLE KNIVES pushes staff away with right knife.

STAFF flows towards left.

Both partners spin away from each other towards their respective left sides.

Both partners assume final position, brandishing weapons.

Conclusion

After preparing the material and executing the photos and the writing for this text, I can only come back to reiterate that the saying that weapons are an extension of the practitioner's hand falls slightly short of the meaning that it tries to convey. Not only is the weapon—in this case, the double butterfly knives—an extension of the practitioner's hand, it is truly an amplification of the mind, body and spirit of the individual wielding these weapons.

It allows one to grow beyond the physical parameters of their body and expand their understanding of the Kung Fu that they have chosen to study, thereby giving them a platform to grow, not only as a Kung Fu practitioner, but as a person. The weapons and the techniques employed flow seamlessly together to develop and amplify the individual's entire understanding of the deep mysteries of the Chinese martial arts.

About the Author

Master Paul Koh is internationally recognized as one of the world's leading Kung Fu teachers. He has dedicated his life to the study of this ancient Chinese art form, immersing himself in Chinese language and culture since his early teens. He is one of the only major advocates for preserving and promoting this cultural art which, despite being frequently overshadowed today by popular fads, is a timeless and ever-evolving art that goes far beyond a punch and kick.

With over 40 years of experience, Master Koh has extensively trained with world-renowned Kung Fu masters in 少林洪家虎鶴拳術 Hung Gar Tiger Crane and 少林黑虎門虎爪派 Fu Jow Pai Tiger Claw Kung Fu. He has expertise in classical Chinese weaponry and in traditional Southern lion dancing. Master Koh is not only a martial artist who has dedicated countless hours, days, months and years to the study of his discipline, he is also a true educator, mentor and motivator. He has taught thousands of students and currently maintains a martial art training hall in NYC Chinatown. His knowledge of Kung Fu goes far beyond the physical execution or application of a technique, and extends to the historical and cultural background of Chinese martial arts, as well as the deep philosophical bedrock upon which this art form depends. These qualities, combined with an ability to speak and write clearly and eloquently, make Master Koh a unique phenomenon in the martial arts world.

In addition to studying, researching and teaching the Chinese martial arts for over four decades, Master Koh is an accomplished author, having written several previous texts on the art of Kung Fu, as well as having published many articles in various martial art magazines throughout the years. He currently serves as martial arts consultant on a mystery novel series published by Soho Press and written by John Shen Yen Nee (former executive at DC Comics and Marvel Comics) and SJ Rozan (author of 18 novels, Edgar Award winner). The first novel, *The Murder of Mr. Ma*, was released in early 2024, and Master Koh's work was hailed by reviewers for his "incredibly orchestrated fight scenes" and "absolutely fabulous displays of martial arts."

In 2018, Master Koh launched Kung Fu In A Minute, a unique approach to documenting, promoting and preserving the traditional skills, attitudes, traditions and practice of the art of Kung Fu. Through Kung Fu In A Minute, he has published 14 titles with more forthcoming. The Kung Fu in a Minute library of videos, articles and manuals will further promote, preserve and protect this ancient art form.

Additional titles available from
Kung Fu In A Minute Publications and Master Koh

Ancient Weapons of Kung Fu Series

虎頭凳
Tiger Head Fighting Bench

Dynamic Art of the Tiger Series

十獨手
Ten Essential Techniques of the Tiger

虎爪木人樁大師
Tiger Claw Wooden Master

八卦虎爪對拆
Tiger Claw Eight Diagram Fighting Set

虎爪八卦棍
Tiger Claw Eight Diagram Long Pole

黑虎爪單刀
Black Tiger Claw Single Saber

虎爪 虎尾三節棍
Black Tiger - Tiger Claw Three-Section Staff

Southern Shaolin Series

嶺南拳術五行拳
Five Element Fist

嶺南拳術仙鶴拳
Immortal Crane Fist

嶺南拳術虎爪手法
Principles of the Tiger

猛虎鐵鎚
Fierce Tiger Iron Hammer

嶺南虎爪雙刀
Tiger Claw Double Knives

虎鶴對拆
Tiger & Crane Matching Set

E-Books

伏地虎尾脚
Tiger Tail Kick: Ground Fighting Technique

黑虎爪對拆
Black Tiger-Tiger Claw Matching Set, vol. 1

For more information, visit
KUNGFUINAMINUTE.COM

Made in the USA
Las Vegas, NV
12 January 2026

35967492R00098